Bromeliad Display at Nani Mau Gardens, Hilo, Hawaii

Kukui Nut, Hawaii State Tree

Printed by:
    Corporate Graphics
    North Mankato, Minnesota

Published & Distributed by:
    Plant Pics
    P. O. Box 3224
    Duluth, MN 55804-3224

Book and cover design by:
    Clayton & Michele Oslund

Cover photo taken at:
    Waimea Arboretum & Botanical Garden

ISBN 0-9667399-0-6

Hawaiian Gardens are to Go to

*A pictorial story of
Hawaiian Gardens based on our experience of
guiding tours through these gardens*

text and photography by

Clayton and Michele Oslund

ISLANDS and GARDENS

Japanese Garden at Honolulu International Airport

Japanese Garden at Honolulu International Airport, lower level

THE ISLAND OF OAHU

Aloha! . . . with a flowery Welcome!

Oahu offers the momentum of Honolulu at the airport as the rush of traffic, both in the air and on the ground, is confirmation that our journey is well underway. With high expectations and the immediate need to collect luggage at the claim area, we hurry away from the lei greeting at the gate. Once reunited with travel gear, we begin to open our senses to this initial island experience by observing Palms, Ti plants, Hibiscus hedges, Pothos vines as groundcovers, and other tropical plants waiting to be identified.

Often the lovely Japanese Garden at the heart of the terminal is overlooked. A walk through this garden is an overture to the many beautiful plantings waiting to be seen throughout the islands.

Honolulu city dresses up every day in garlands of flowering plants. Cascades of Bougainvillea vines hang above our heads from planter boxes on the facades of hotels and shopping centers. Whether lavish or simple, neatly trimmed lawn areas are landscaped to catch the eye and, just as much, the soul. Always, the rustle of the ever-stately palms reminds us that we are in the Tropics.

Hedges of Hibiscus define many streets as do masses of green and red Ti. Ti plants vary in size and color, but all are important in the belief that they bestow good fortune on the people who use them.

Hibiscus flowers last for a single day; then the blossom wilts and dies. It is permissible to pick a Hibiscus blossom, but all other temptations to gather bouquets should be squelched.

Driving along Ala Moana Boulevard from the airport to Waikiki Beach, we see these beautifully sculpted trees in the Ward Center vicinity.

HONOLULU CITY

Queen Kapiolani Garden

Tucked away behind the Honolulu Zoo, the garden waits to show off royal collections of hybrid Hibiscus, Crotons and Bougainvillea.

An arbor of pruned Hau, or Hawaiian Hibiscus, grows above our heads like a crown. Along with sea breezes wafting in from the ocean over Waikiki Beach, the shade under the Hau arbor is refreshing on a warm day.

From the corner of Monsarrat and Paki Avenues, turn left to the main entrance and parking area one block from the intersection.

Dwarf Poinciana (Caesalpinia pulcherrima)

At the edge of the parking lot, Dwarf Poinciana trees give a show of color, both in orange-red and yellow clusters.

Dwarf Poinciana and Monkey Pod Trees are in the legume (pea) family.

Large trees spread their generous canopies like umbrellas over the land. These giants are usually dormant in March, but their flowers give them away in springtime. The Monkey Pod tree blossoms look like pink and white "puffs" in bloom.

Noni, *Morinda citrifolia*, is a member of the coffee family, but its fruit does not taste like coffee. Although the fruit can be eaten and will sustain life, it has an unpleasant taste.

Loulu Palm, *Pritchardia remota*, is native to the Hawaiian islands and is often used as a landscape Palm.

Like elegant embroidery at the neck of a royal robe, Croton plants, *Codiaenum variegatum,* face into the sun along the pathway. Seeing these colorful foliages used in landscape settings, cold climate gardeners may turn green with envy!

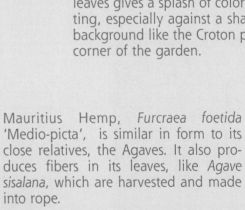

Snowbush, *Breynia nivosa,* is typically grown on the islands as a hedge material. White and pink variegation of the leaves gives a splash of color to any setting, especially against a shady or dark background like the Croton plant in this corner of the garden.

Mauritius Hemp, *Furcraea foetida* 'Medio-picta', is similar in form to its close relatives, the Agaves. It also produces fibers in its leaves, like *Agave sisalana,* which are harvested and made into rope.

Looking toward the zoo, mature canopies of Monkey Pod trees, *Albizia saman,* stand like giant umbrellas against the sky. Flowers are shown below.

4

'Burr Philips'

'Queen Elizabeth'

'Jason '

Hybrid Hibiscus Cultivars

Hibiscus is the state flower of Hawaii. This garden includes dozens of hybrid cultivars which are ablaze with hundreds of blossoms in the warmer seasons! The blooms shown are a small sampling of the cultivars and colors.

Even though the Islanders tell us it is permitted to pick and wear a Hibiscus flower, these beauties are best left for other visitors to enjoy.

'Lunar Dust'

'Keahu'

'All Aglow'

Just inside the entrance of the garden, across from the Croton collection, a tree-form Bougainvillea gathers sunlight to dazzle us with its display of richly-tinted bracts. Bracts are modified leaves showing color. (Poinsettias are other plants having colored bracts that resemble flower petals.)

'Orange Splash'

Bougainvillea Variations

The actual flowers of Bougainvillea, *Bougainvillea sp.*, are small and white, in groups of three. Finding all three white blossoms in the bracts open at the same time is a challenge.

The Bougainvillea archway below illustrates the versatility of this popular landscape vine.

6

Honolulu Botanical Gardens

Honolulu Botanical Gardens are a series of five gardens located throughout Oahu. Each garden is found in a unique ecological setting and displays plants from the common tropical and subtropical to rare and unusual specimens collected from around the tropical world.

Foster and Lili'uokalani Botanical Gardens are located in bustling Honolulu, separated by the H-1 Freeway! They were once one garden, the Foster Estate, but modern development divided them. The ecological setting provides ideal conditions for plants of a sub-tropical climate.

Wahiawa Botanical Garden is near the center of Oahu in the town of Wahiawa. The elevation of one-thousand feet provides a location for plants from cool, humid tropical climates.

Koko Crater Botanical Garden is on a hot, dry site in south-eastern Oahu, exhibiting xerophytic plants that are adapted for hot areas of low rainfall.

Ho'omaluhia Botanical Garden in the high rainfall area of Oahu has gardens within the garden providing unique collections of plants from specific tropical regions: Africa, Hawaii, Malaysia, Melanesia, Philippines, Polynesia, Sri Lanka/India, and Tropical America.

Foster Botanical Garden

Easily reached by bus or driving, this former estate is located at 50 North Vineyard Boulevard. Queen Kalama leased this property to a young German physician-botanist, Dr. William Hillebrand, and his wife in 1853. During their twenty year stay, Dr. Hillebrand planted trees which now tower majestically overhead. Today these trees are considered rare and are labeled "exceptional," a status indicating historical or cultural significance, given by the County Arborist Committee as worthy of preservation. Thomas and Mary Foster became the next owners of the house and its now "growing" botanical collections. Mrs. Foster continued the development of the grounds until her death in 1930.

Gold tree
(Tabebuia donnell-smithii)

Termites destroyed the wooden structure of the house and in time, the building was removed. Happily, we still have the gardens to explore! Included in Foster's collections is an economic garden of herbs, spices, dyes, and other plants that provide products of commerce. Take time to enjoy a gathering of Palms and the "prehistoric" garden of Ferns and Cycads.

A nominal admission fee allows a visitor to stroll through the garden on a self-guided tour assisted by well-prepared pamphlets and maps. Docents are ready to guide groups, but please schedule a guided tour in advance. Phone: 808-522-7060.

Lifting leafless branches to the sky, the Silk Cotton Tree is not a lifeless skeleton. It is dormant. For this tree, winter has arrived. Like temperate zone deciduous trees, it has shed its leaves in preparation for new growth in the spring. Deep-red blossoms break out of their buds beginning in early spring. The blossoms develop into large capsules filled with white, silky strands, soft to the touch. The Silk Cotton Tree is related to the Kapok Tree whose fibers were used to stuff life-jackets.

The giant Quipo Tree, *Cavanillesia plantanifolia*, stretches above the towering palms. It is a fast growing tree with a soft wood, softer than the well-known Balsa wood used in making model airplanes.

Silk Cotton tree
(Bombax ceiba)

An "exceptional" tree label is displayed on this Chicle Tree. Chicle fruit is delicious and the sap is a major ingredient in chewing gum.

The "exceptional" Royal Palm, *Roystonea oleracea*, stretches toward the sky as it extends up through and beyond the Chinese Banyan Tree (*Ficus microcarpa*). The Banyan's hanging roots have touched the ground and have become supporting trunks for the parent tree.

Queensland, Australia, is the native home of the Kauri Tree, *Agathis robusta,* a broad-leaved conifer. This giant is truly an "exceptional" tree in the heart of Honolulu.

Photographed in March, the Kapok tree, *Ceiba pentandra*, (left) was beginning to grow a new crop of leaves. Kapok trees produce a waterproof floss within their seed capsules, good for stuffing life preservers and other padded furnishings. The picture below was taken in August when the tree is in full leaf.

Changing seasons are displayed in tropical trees and shrubs, varying from island to island and coastline to mountaintop.

"Dogbones" from the Dogbone Tree, *Polyscias nodosa*: Large leaves have jointed petioles that break apart in sections resembling "bones" when the leaves drop from the tree. Look up to see a large canopy of dark green leaves atop a pole-like tree when you find these on the pathway.

Small peppercorns form on the high climbing Black Pepper Vine, *Piper nigra*. Both black and white pepper are produced from these peppercorns.

Cannon-ball Tree *(Couroupita guianensis)*

Large round fruits, "cannon-balls," hanging in clusters on the trunk, give the tree its name. Striking color in the flowers adds unusual beauty to this tropical tree.

Chocolate and coffee trees are grown in the shade of the "Mother of Chocolate" Tree in some areas of the tropics. Pea-shaped flowers are a telltale indication that this tree belongs to the legume family.

Mother of Chocolate *(Gliricidia sepium)*

Golden Shower, Saraca and Brownea are examples of trees in the legume family. Legume trees are abundant in the tropics.

Baobab trees are native to Central Africa. Sometimes they are referred to as "dead rat" trees because their fuzzy fruits hang from long stalks resembling rat tails. Baobabs adjust their own growth schedule: When each tree decides it is tall enough, it stops growing up and begins to grow in width. In their native land, large trunks are hollowed out for use as shops and living quarters.

Golden Shower Tree *(Cassia fistula)*

Baobab or Dead Rat Tree *(Adansonia digitata)*

Brownea *(Brownea macrophylla)*

Red Saraca *(Saraca declata)*

Lili'uokalani Botanical Garden

Located across the H-1 freeway from Foster Botanical Garden, Lili'uokalani provides an oasis of quiet in the neighborhood. Flanked by large apartment buildings, one of its unique characteristics is a collection of native Hawaiian plants. And a river runs through it.

Mao or Hawaiian Cotton (Gossypium tomentosum)

From Foster Botanical Garden, Lili'uokalani is easy to reach by turning right onto Vineyard from Foster's parking lot. Turn right on Liliha Street and continue on to Kaukini, making another right turn. Go past the hospitals. The garden entrance is on the right (ocean-side) of the street.

Wiliwili or Hawaiian Coral Tree (Erythrina sandwicensis)

Scarlet Bottlebrush
(Callistemon citrinus)

Ho'omaluhia Botanical Garden

Four hundred acres of tropical beauty, Ho'omaluhia is a flood control project built by the U. S. Army Corps of Engineers, including a thirty-two acre "lake" or reservoir. A perfect blend of gardening skills and engineering science, the garden spotlights plantings from specific geographical regions. Africa, Hawaii, Malaysia, Melanesia, Philippines, Polynesia, Sri Lanka/India, and Tropical America are represented here. Paved roads and hiking trails guide the way to each region. The plants are well-labeled and accessible along walkways and in mowed alleys.

Located in a high rainfall area on the windward side of Oahu, follow either the Like-Like (sounds like "leaky-leaky") Highway or the Pali Highway. The Pali Highway is our suggested route: by stopping at the Nu'uanu Pali Lookout, we get a bird's-eye view of the garden settled between the mountains and the city of Kaneohe. From the Pali Highway, turn left on Kamehameha Highway. Look for Luluku road, turn left again and follow the signs. Stop at the visitor center for maps and information.

The name Ho'omaluhia, means *"to make a place of peace and tranquility."*

Inside Ho'omaluhia

Flame Trees, *Brachychiton acerifolius,* native to Australia, have large glossy leaves that remind us of Maple leaves. Brilliant red bell-shaped flowers add a delicate, lacy look.

Royal Poinciana, *Delonix regia,* is rarely found in this yellow form. A distant cousin, Dwarf Poinciana, also gives a show of sunny-yellow blossoms.

The White Orchid Tree is related to the commonly seen Hong Kong Orchid Tree. Only their flowers resemble orchids. Butterfly-shaped leaves add an exotic touch.

White Orchid Tree
(Bauhinia variegata 'Candida')

Pickle and Lipstick trees are curiosities. The "pickle" fruits can be eaten but are truly sour. Seeds inside the prickly red fruits of the Lipstick Tree contain a dye that is used for coloring food and cosmetic products.

Lipstick Tree
(Bixa orellana)

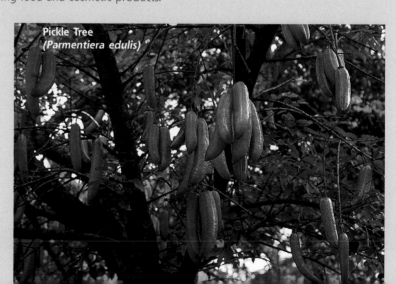

Pickle Tree
(Parmentiera edulis)

Lonchocarpus sericeus

Osmoxylon linearea

Ten-cent Flower Tree
(Fagraea berteriana)

Golden Ohi'a lehua(Metrosideros polylmorpha)

Lonchocarpus and Osmoxylon above are attention getters in Ho'omaluhia. Golden 'Ohi'a lehua grows well in the rainy climates of the islands. White 'Ohi'a lehua Trees happen but are rare.

Kaffirboom Coral Tree is an Erythrina, one of many species found in the tropics, all of which are legumes.

Named for the tradition of selling the fragrant flowers for ten cents each, used for making leis, lovely white flowers soon turn to gold.

Kaffirboom Coral Tree
(Erythrina caffra)

Giant Crepe Myrtle
(Lagerstromia speciosa)

16

Not the usual Plumeria, but a dainty flowered species, *Plumeria stenophylla* is rarely found outside a botanical garden in Hawaii.

Widely used landscape Plumerias (*Plumeria rubra*) are often called Frangipani.

Hawaiian Schefflera, *Schefflera arboricola*, a common houseplant, seldom produces flowers and fruit indoors.

Gardenia volkensii pictured below, another example of a gardenia species, grows in tropical regions. New flowers begin white and become yellow as they age. The fruits look like small cantaloupes, but please don't eat them.

White Mussaenda, *Mussaenda philippica*, bracts reflect the lush green rain-forest growth around them. Its small orange blossom stands at attention above the shadows.

Striking scarlet flowers cluster along this Indian Almond Vine, especially colorful set against the dark foliage. Interesting fruits add highlights to this handsome plant.

Indian Almond *(Poivrea cocciena)*

Adam's Apple fruit *(Tabernaemontana crassa)*

Adam's Apple flower

The Adam's Apple Tree is in the same family of plants as Plumeria and Oleander.

Variegated Hau
(Hibiscus tiliaceus)

Hau trees have variegated or solid green leaves; variegation is more unusual. Hawaiian Hau is another Hibiscus species growing in the tropics.

Hau flower

18

Nearly at the center of Oahu, Wahiawa town embraces twenty-seven acres of garden. Look for the entrance on California Avenue by Nanea Street.

There are good hiking trails here. Plants thrive at this one-thousand foot elevation, and so do mosquitoes: Bring repellant and/or afterbite remedies. Plan a little extra time to visit this garden; its collections include Native Hawaiian plants and other specimens such as the Moreton Bay Fig, Elephant Apple Tree, Earpod Tree and Queensland Kauri.

Along the way . . . Pineapple family plants (Bromeliads) are displayed at the Dole and Delmonte trial gardens just north of town.

Wahiawa Botanical Garden

Begonias grow large on the islands and Wahiawa is no exception.

Begonia sp.

Philodendron Vine *(Philodendron sp)*

Philodendron, a common household word, is the genus name for a large number of tropical vines. Philodendrons are grown as houseplants in and away from the tropics.

The Swiss Cheese Plant, often called Split-leaf Philodendron, is really in the genus *Monstera*. Its scientific name is *Monstera deliciosa*. When grown in its tropical habitat, it produces edible, cone-shaped fruits which are considered a delicacy.

Swiss Cheese Plant *(Monstera deliciosa)* with fruits

Candle Tree flowers and bean-shaped seed pods grow directly from the trunk and large, older branches. This is a typical growth habit of trees in the tropics. Blossoms face down toward the ground, hiding their beauty. Sneak a peak anyway.

Candle Tree
(Parmentiera cereifera)

Mule's Foot Fern fiddlehead *(Angiopteris evecta)*

Fern "leaves" are properly called fronds. These newly emerging fronds look like the scroll at the end of a violin, the "fiddlehead." Mule's Foot Fern gets its name from the way the frond scar looks at the base of the plant after the frond dies and falls off.

Eucalyptus trees can be found all over tropical regions. This "exceptional" Mindanao Gum Tree, *Eucalyptus deglupta*, is one species of Eucalyptus. As the outer layers of bark peel away, this tree shows how it could be called a "Painted" Eucalyptus.

Koko Crater Botanical Garden

Unidentified Cactus in the basin of Koko Crater.

Petra, *Petrea arborea*, grows among the dryland species.

Firecracker Cactus
(*Cleistocactus
smaragdiflorus*)

Yucca sp.

Opuntia sp.

A garden under development, it is also a garden of surprises! Being the interior of an extinct volcano, the hot, dry landscape is an adventure by itself. Enjoy the scenery as the ancient sides of the cone rise up and make a unique backdrop for the collection of Cactus plants found here. Other drought-resistant plants are displayed as examples of xerophytes.

The garden is located in the crater at 400 Kealahou Street. Follow the sign for Koko Crater Stables.

In contrast, by using irrigation and other water conservation techniques, a lush grove of Plumeria trees is thriving, saturating the air with fragrance in bloom-time! From lush Plumerias to arid conditions in which Cactus flowers startle us with their intense color, there is something in this garden for everyone.

Chadsia grevei

Yucca sp.

Unnamed cultivar of *Plumeria rubra*

'Kauka Wilder'

Hybrid Plumeria Cultivars

'Lei Rainbow'

'Irma Bryan'

'Mary Moragne'

During the spring season, multitudes of blossoms decorate the hybrid Plumeria cultivars in the grove near the entrance of the Koko Crater garden.

In other locations, the fragrant flowers are harvested each morning to be used in lei-making. A full lei of Plumeria flowers stays sweet-smelling all day.

Inspiration Point, a look back at the Valley from the Mountainside

Described by John Waihee, Governor of Hawaii, in 1992 as "an Eden in our Urban Backyard," Harold Lyon Arboretum is a unit of the University of Hawaii. Having a mission of research, instruction, and public service, the arboretum boasts of having a collection of over four thousand species and cultivars of tropical plants. Especially important in their mission is the restoration of the Hawaiian ecosystems. Harold Lyon Arboretum is located in the Manoa Valley and is reached by driving to the end of Manoa Road. It is helpful to call ahead (800- 988-7378) to make arrangements for a visit. A donation of $10 per person to the Lyon Arboretum Association is suggested for a guided tour.

Although certain areas of the arboretum are used for research and are not available for public access, nearly eighty acres are planted with native Hawaiian and exotic plant species. A tour through the Beatrice H. Krauss Ethnobotanical Garden adds insight into the lives of early Polynesians. These plants were used to sustain life, heal, and become works of art by the ancients. Many are still used and appreciated as old crafts and arts are taught today.

Well worth the time to see are the herb and spice gardens and the palms growing here, one of the largest palm collections on the Islands.

Mussaenda 'Dona Luz'
(Mussaenda erthrophylla)

Mussaenda is best seen in the fall season on the islands. Colorful red-pink bracts support small flowers inside the bract cluster. The white form is equally welcome in the garden landscape. Often, these colorful shrubs grow to the height of a small tree.

Chocolate lovers may be surprised to learn that the two components of cocoa fruit used commercially, the cocoa butter and the substance familiar to us as chocolate, are actually white during processing. The dark color is added. Cocoa fruits ripen to the rich red seen in the photograph on the right. Dried husks are ground for use as garden mulch, especially pleasing because of its chocolate fragrance. Typical tropical growth shows the blossoms and the fruit growing directly from the trunk and branches.

Cocoa Tree blossoms   Cocoa Tree *(Theobroma cacao)* fruit

Cloves (below) also ripen red. Then the buds are picked, fermented, and dried before they reach our spice racks.

Syzygium aromaticum

Nutmeg and Mace are extracted from the seed of the peach-like fruit below.

Myristica fragrans

Tea *(Camellia sinensis)*

Real Tea comes from this plant closely related to Camellias and should not be confused with red and green Ti plants which grow throughout the islands.

Unusual houseplants, different because they form a vase of water at the base of the leaves, are in the Pineapple family, *Bromeliaceae*, commonly called Bromeliads. When these plants are ready to start blooming, side shoots may appear and grow as "babies." The mother plant then puts up a flower stalk, often producing brightly colored bracts that stay "in bloom" for many days or weeks.

Once a Bromeliad has finished its bloom-time, the mother plant dies and the babies continue the cycle of growth. Some of the Bromeliads commonly grown as houseplants are Guzmania and Tillandsia. The Tillandsias are fun because, growing from tree branches, they get their water supply from humid air, giving them the nickname, "Air Plants."

Queen Aechmea
(Aechmea mariae-reginae)

Aechmea aqueliga

Giant Tree Fern Forest in Harold Lyon Arboretum

Waimea Arboretum and Botanical Garden

Travel to the north shore of Oahu to find Waimea Valley. Not simply a garden, this park provides a restaurant, a gift shop, a small zoo, Hawaiian cultural demonstrations (such as Hula lessons), and from sixty feet, near the top of the waterfall, cliff diving. The cliff diving performances are done only by professionals, but visitors are invited to swim in the pool.

The gardens encompass thirty-six collections of tropical plants, naming origins from nearly that many tropical regions. Plan at least a half a day to get a proper perspective on all the flora displayed and to take in the other sights and demonstrations. With about one and one half miles of paved walkways, a visitor is invited to take leisurely hikes through the garden areas or ride the trams. Trams usually have a guide who can give some timely explanations about the plants and places of interest along the way. Admission is charged. For information call 808-638-8511.

Tamarind

Thunbergia

Tamarind *(Tamarindus indica)*, a tropical evergreen tree, is valued for its gracefulness in the landscape. It is in the legume family. The Thunbergia Vine *(Thunbergia mysorensis)*, adds lively color wherever it may hang.

Coconut Palm (Cocos nucifera) and Royal Poinciana (Delonix regia)

Variegated Alocasia *(Alocasia macrorrhiza)*

Waimea Falls , site of Diving Performances

Yellow Cape Honeysuckle *(Tecomaria capensis 'Aurea')*

Divers display their skills here at the waterfall. During rain storms up on the mountains, this river runs full. In drought conditions, there is little more than a trickle. Water from the mountains is cool and clear. The green cast on the water of the pool is the reflection from the mountainside.

Monstera deliciosa, or Swiss Cheese Plant, becomes ground cover in the arboretum.

Hanging Heliconia (*Heliconia collinsiana*)

Hanging Heliconia
(*Heliconia rostrata*)

Lobster Claw (*Heliconia bourgaeana*)

Heliconia Variations

Beautiful clusters of bracts may stand firmly upright or gracefully hang down in the Heliconia collection. Common names such as "Lobster Claw" and "Parrot's Beak" help to define and describe these showy plants.

Handsome Heliconias are in a family all their own, *Heliconaceae*. An older classification system put them in the *Musaceae* or banana family. Leaves of Heliconia resemble those of bananas.

Red-Leaved Heliconia
(*Heliconia indica* 'Spectabilis')

Parrot's Beak Heliconia
(*Heliconia psittacorum*)

Small Lobster Claw Heliconia (*Heliconia stricta*)

Heliconia (*Heliconia lingulata*)

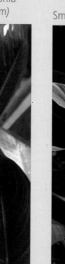

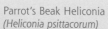

Orange Latispathas Heliconia *(Heliconia latispathas)*

Giant Lobster Claw Heliconia *(Heliconia caribaea)*

Red Flame *(Hemigraphis alternata)*

Coral Aphelandra *(Aphelandra sinclairana)*

Red Flame and Coral Aphelandra are in the Acanthus family, a large and variable group of plants. Characteristics used in naming plants of the Acanthus family are confusing. For example, one may see the genus *Ruellia* and *Hemigraphis* applied to the same plant.

Caricature, Shrimp, Zebra, and Silver-vein plants and more in the Acanthus family are represented at Waimea.

Certain Aphelandra species are sought after as houseplants for their pretty foliage as well as their flowers. One common household name is "Zebra Plant" *(Aphelandra squarrosa)*.

This Coral Aphelandra is exploding with color as its flowers poke out from fingers of bracts.

30

Costus (*Costus claviger*)

Hidden Lily (*Curcuma ornata*)

White or Butterfly Ginger (*Hedychium coronarium*)

Members of the Ginger family, Costus, Curcuma, and Ginger plants are grown for their ornamental value. From small to tall, these tropicals can color any garden spot with beauty.

The spice Ginger, *Zingiber officinale*, is commercially harvested from its underground stems (rhizomes). Called Ginger "Root," it is not root at all.

Another spice plant, *Curcuma domestica* is harvested for its rhizomes as the spice turmeric, used in the making of turmeric pickles. A dye product is also extracted from this species.

Spiral Flag (*Costus speciosus*)

Pink Ginger (*Alpinia purpurata* 'Jungle Queen')

Varigated Costus (*Costus amazonicus*)

Where mountain meets lawn on the way down to the valley floor, a resting place spreads out like a welcome mat. A good place to picnic, soak up sun, or cool off in the shade, a visitor can take a moment to enjoy the view.

Blue ginger, *Dichorisandra thysiflora*, is in the Spiderwort family along with Tradescantia and Zebrina. It is not a "ginger." This is one of many misnomers in naming plants.

Bauhinia shrubs and trees look exotic when not in bloom; their unusual leaves may resemble a Valentine's Day heart or butterfly wings. One small tree, commonly called the Hong Kong Orchid, has an eye-catching pink to lavender flower. Equally lovely is the White Bauhinia. Here at Waimea, Red Bauhinia, *Bauhinia punctata*, is a stand-out.

Varigated Pineapple *(Ananas comosus)*

More Bromeliads!

All of these being related to the pineapple, it is not a surprise to find a decorative pineapple used as a garden display.

"Little Basket" Bromeliad *(Canistrum aurantiacum)*

Bromeliad *(Hohenbergia stellata)*

Aechmea Bromeliad *(Aechmea blanchetiana)*

Pink Quill Bromeliad *(Tillandsia cyanea)*

Delmonte and Dole maintain demonstration plots of wild and cultivated pineapple species and varieties. The Delmonte garden is a short distance north of Wahiawa in the center of Oahu at the junction of Highways 80 and 99.

The Dole display, at the Dole Pineapple Pavilion is just a short drive further north. Both gardens are well labeled to illustrate the origin of pineapple and pineapple production.

False Heather, which is a common groundcover in subtropical climates, is found at the Dole Pavilion.

Night Blooming Cereus can be seen (if you get up early in the morning!) on the cliffs along the road to Punchbowl Crater where the National Cemetery of the Pacific is located. An overlook with a view of Honolulu and a well-maintained garden are inside the crater.

Ornamental Pineapple
(Ananas ananassoides)

Night Blooming Cereus *(Hylocereus undulata)*

Garden Patches on Oahu

Wild-growing False Koa, *Leucaena leucocephala*, covers steep hillsides, roadsides, and undeveloped areas.

False Heather *(Cuphea hyssopifolia)*

Flower of Love
(Tabernaemontana divaricata)
at Byodo-In

Giant White Bird-of-Paradise, *Strelitzia nicolai*, are found at the Arizona Memorial grounds.

The Byodo-In is a Budhist temple in a peaceful garden setting on the windward side of Oahu. An Oriental-style garden welcomes travelers, and the Koi in the large ponds are very entertaining.

34

Mauna Kea Mountain rises above Hilo Bay

The Island of Hawaii

Called the "Big Island," Hawaii is rich in botanical diversity, holding treasures at every turn. From the wet, windward side, to the stark volcanic regions (some are just beginning to develop soil and vegetation), toward vast grassland areas on the drier, leeward side of the mountains, the plants growing here certainly contribute to the life-styles of this island's inhabitants.

In the development of the island's agriculture, crop production has left its marks in history. In decades past, sugar cane fields meant thriving economy. More recently, Macadamia Nut and Coffee Trees take their place. Orchids and other exotic species grown for the cut-flower trade are important crops. Ranching is big business, especially at the Big Island's Waimea area.

Always holding the land away from the sea, like a mother lifting her child from the water, the mountains Mauna Loa and Mauna Kea remind us how close they are; yet their peaks remain ever distant and majestic, especially in winter, wearing crowns of snow.

Lili'uokalani Park is kept in traditional Japanese form. This timeless garden with its pools, rock formations, and bridges, is a place of serenity in busy Hilo.

Lili'uokalani Park at Hilo

Topiary Trees on Lava Rock Formations near Coconut Island

Hawaii Tropical Botanical Garden

Described as a tropical nature preserve, the Hawaii Tropical Botanical Garden is a seventeen-acre site in Onomea Valley on Onomea Bay, just seven miles northeast of Hilo. The word onomea means "the best place" in Hawaiian, and many visitors to the garden agree. Over 2,000 species representing more than 750 genera and 125 families of plants can be enjoyed and studied here.

Each trail brings the visitor to unexpected moments of beauty as plants are displayed along the pathways and in garden "rooms." Views of the ocean add visual flavor; the starkness and might of surging waters contrast with the lush garden anchored in the valley. From the mountain side of the valley, Onomea Stream tumbles headlong down its rocky waterway in a rush to reach the sea.

An admission fee is charged. Membership in the garden entitles one to free admission, a newsletter update of garden happenings, and a discount at the book and gift shop.

Onnonea Falls at Hawaii Tropical Botanical Garden

Persian Shield *(Strobilanthes dyerianus)*

Bat Flower *(Tacca chantrieri)*

Beehive Ginger *(Zingiber spectabile)*

Costus *(Costus curvibracteatus)*

Alakahi Stream

Giant Spiral Ginger *(Tapeinochilus ananassae)*

Rainbow Plants (Dracaena tricolor) Glow in the Sun

Rattlesnake Plant
(Calathea insignsis)

Segmented bracts may remind us of "rattles," but there is nothing worrisome about this Calathea. It will not bite! A member of the Arrowroot family *(Marantaceae)*, it has cousins that are crop plants, producing arrowroot starch and leaves that are processed for wax.

Jakfruit *(Artocarpus heterophylla)*

Calathea blossom *(Calathea sp.)*

Multi-colored Ti (Cordyline terminalis) Brightens the Pathway

Akaka Falls State Park

Giant gingers, bamboo groves, and tall trees draped with living cloaks of vines wait to greet visitors on the hiking trail that loops past both Kahuna Falls at 400 feet and Akaka Falls at 420 feet. Kukui Nut Trees, their silver-green foliage making them easy to identify, cover the hillsides.

Akaka Falls

Swiss Cheese Plant (*Monstera deliciosa*)

Cup of Gold (*Solandra maxima*)

Ficus tree aerial roots

Yellow Bamboo
(Phyllostachys sp.)

Pink Banana *(Musa velutina)*

Nani Mau Gardens

In the Hawaiian language, the words Nani Mau mean "forever beautiful," a description truly fitting this garden. Gathered here are collections of Anthuriums, Bromeliads, Gingers, Heliconias, Orchids, tropical fruit and nut trees, Water Lilies -- and more!

An admission fee is charged. Stroll through the lanes on a self-guided tour, join a guided group, or motor through the garden on a tram. Explore the botanical museum, gift shop, and restaurant.

From the Hilo Airport intersection on Highway 11, travel 2.9 miles toward Volcano; turn left on Makalika street.

This tree fern pathway demonstrates a traditional method of growing Anthuriums for cut-flower production. Today tree ferns are replaced by shade cloth in commercial nurseries.

Hawaiian Tree Ferns *(Cibotium sp.)*; Anthuriums *(Anthurium spp.)*

Pink Tecoma *(Tabebuia rosea)*

Surinam Cherry blossom *(Eugenia uniflora)*

"Forever Beautiful"

Shaving Brush Tree *(Pseudobombax elipticum)*

Panama Flame Tree *(Brownea macrophylla)*

Pink Quill Tillandsia *(Tillandsia cyanea)*

Tillandsia bracts take interesting forms and show vivid colors. Pink Quill Tillandsia bracts have a feather-like appearance which contrast with its vivid blue flower.

Brilliant red is spotlighted against deep green, perhaps reminding us of "Christmas" all year long on the islands. Like many other bromeliads, Guzmanias make handsome houseplants.

Zamia pumila (It's such fun to say!) is commonly called "Cardboard Palm," but it is really a Cycad. Cycads have remained relatively unchanged through countless ages. Fossil records confirm their ancient existence.

Cardboard Plant
(Zamia pumila)

Guzmania Bromeliad *(Guzmania lingulata)*

Pink Quill Tillandsia
(Tillandsia cyanea)

New Guinea Creeper *(Tecomanthe dendrophylla)*

Torch Ginger *(Etlingera elatior)*

Shell Ginger *(Alpinia zerumbet)*

Indian Head Ginger *(Costus spicatus)*

Giant Aechmea Bromeliad *(Aechmea sp.)*

Tillandsia sp.

Vriesea sp.

Sausage Tree and Blossom (Kigelia pinnata) in Nani Mau Gardens

Fern Palm, *Cycas circinalis*, is neither a fern nor a palm. It is a cycad bearing cones with seeds produced on scales similar to that of pine trees.

Queen Tree (*Amherstia nobilis*)

Lava Tree State Park

Ohi'a Lehua (flower)

Volcanic eruptions bring permanent changes to the landscape. Two centuries ago, lava flowed down the mountainside and devastated this forest. Larger Ohi'a Lehua trees caused the lava to harden around them before the tree was consumed. As the flow of lava drained away, these stone "trees" remained.

Lava Trees

Ohi'a Lehua (Metrosideros polymorpha)

Heading south of Hilo, take Highway 130 at Kea'au; continue on to Pahoa, turn left on Highway 132 for three miles. Look for the entrance on the left. Enjoy the wild Impatiens along the roadsides.

Lava tree with Pandanus (Pandanus tectorius)

Ohi'a Lehua (flower)

'Ohi'a Lehua Tree seeds germinate in the sterile soil found on lava fields allowing them to be one of the first plants to colonize old lava flows.

Terrestrial Bambo Orchids, *Arundina bambusaefolia*, grow among False Staghorn Ferns, *Dicranopteris linearis*.

Philippine Ground Orchid *(Spathoglottis plicata)*

Volcanoes National Park

Visiting Volcanoes National Park is like being transported back in geologic time. From the barren rock of new lava flows to the old flows that steadily collect soil, progressions of plant life can be seen filling the landscapes.

Trails at the Thurston Lava Tube are closed in by giant tree ferns standing tall, noiselessly unfurling their velvety fiddleheads. Among the ferns, Ohia trees and wild Fuchsias thrive at this sinkhole.

Tree Fern Fiddleheads

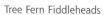

Hawaiian Tree Ferns *(Cibotium spp.)*

Wickstroemea uva-ursi

Sadie Seymour Botanical Garden

Plant collections from different geographical regions are beautifully presented and maintained in this contained but dynamic garden! Terraced lawn areas invite a leisurely stroll. Please take time to explore: An ancient heiau keeps its secrets just beyond a garden wall.

Sealing Wax Palm *(Cyrtostachys renda)*

Located in Kailua-Kona at the junction of Kuakini Highway and Highway 11, arrive on Kuakini Highway from downtown Kailua. Watch for the stone wall on the right with a gate opening just before reaching Highway 11.

Ixora *(Ixora chinensis)*

50

Ruellia affinis

Abutilon menziesii

Water Hyacinth *(Eichhornia crassipes)*

Hibiscus arnottianus

Indonesian Jatropha *(Jatropha gossypifolia)*

*Palm Glade at
Sadie Seymour
Botanical Garden*

Puuopelu, once home to John Palmer Parker II

Parker Ranch Historic Homes and Gardens

Furnishings in the homes are examples of hand-carved Koa wood, such as the beds adorned with Hawaiian-style quilts in Mana Hale (original home of John Palmer Parker) and fine European heirloom pieces at Puuopelu, their graceful style befitting royal grandeur, significant in the early days of Parker Ranch and perpetuated by Richard Palmer Smart (1913 -1992), sixth generation of the Big Island lineage of Parkers.

Entrance Driveway to the Historic Homes & Gardens

At this higher elevation, temperate zone plants thrive among tropicals. The rose garden is a charming surprise.

Rose (Rosa sp.)

Rose *(Rosa sp.)*

High in the grassland ranges, Paniolos (Hawaiian cowboys) continue a ranching style that began a century and a half ago when John Palmer Parker became the owner of two acres of land a few miles southwest of Waimea (Kamuela) town.

Years before, cattle had been shipped to the island, but without herdsmen, these animals roamed wild, slowly but steadily destroying native habitat. Finally brought under control by John Palmer, the wild herds became a unique cattle industry and tradition at Waimea.

Swiss Cheese Plant, *Monstera deliciosa*, covers a stone wall leading to the planned garden of Native Hawaiian plants.

Neatly clipped hedges define a formal outdoor room.

Jacaranda *(Jacaranda obtusifolia)*

In full springtime bloom, Cup of Gold Vine *(Solandra maxima)* grows as a living crown over this statuaried memorial.

Richard Smart, accomplished actor and singer, began collecting French impressionist art in 1951 when he was a headline performer at the famed Lido Club in Paris, France.

His choice of fine art included Peking glass from the nineteenth century and exquisite glass chandeliers from Italy. A pre-Tang Dynasty ceramic horse holds stage center in the spacious living room of Puuopelu.

Magnolia *(Magnolia x soulangeana)*

Gardenia *(Gardenia jasminoides)*

Amy Greenwell Ethnobotanical Garden

Now owned by the Bishop Museum, this land was once worked as a coffee plantation for commercial production. This historic garden is an upland farming site once used by Hawaiian farmers to support a large population in the years before Captain Cook sailed to Hawaii.

These fields were planted with a wide variety of crops including taro, breadfruit, sugar cane, ti, sweet potatoes and yams.

Long, low ridges of black lava rock reach down the hillside like fingers from a giant hand gently caressing the land. These ridges are remnants of the Kona Field System, a network of gardens tended by skillful Polynesian farmers.

Prickly Poppy
(Argemone glauca)

Taro has been a staple food crop on the islands for generations. All parts of the taro plant are edible. Its fleshy,
underground corm can be baked like potato or mashed
and mixed with water into a paste called *poi*. Grown both
in water or on dry land, taro is becoming popular as a
decorative garden specimen as well as a food source.

Purple and Variegated Taro *(Colocasia esculenta)*

'Ohai *(Sesbania tomentosa)*

Kohi'o *(Kokia drynarioides)*

Sugar cane takes its place in the proud circle of agricultural crop plants the early Polynesians brought with them in their canoes as they sailed the ocean.

Chewing on the sweet fibers was the island way of encouraging children to "brush" their teeth and promote healthy gums.

Sugar cane is propagated by cutting the stem into "planting pieces" called pula-pula.

Sugar Cane *(Saccharum officianarum)*

New shoots forming at the nodes of the sugar cane stem.

Uhi or Yam *(Dioscorea alata)*

Uhi' Uhi *(Caesalpinia kauaiensis)*

Dioscorea alata is a true yam, not to be confused with sweet potatoes, which are often marketed as "yams." The flesh of the true yam is usually white.

Yams, *Dioscorea*, are in the *Dioscoreaceae* family.

Sweet potatoes, *Ipomoea batatas*, are in the *Convolvulaceae* or Morning Glory family.

Garden Patches, Island of Hawaii

Floral Display at Akatsuka Orchid Gardens, Volcano

Macadamia Nut Tree *(Macadamia integrifolia)*

Coffee *(Coffea arabica)* berries

Red and Gold African Tulip trees *(Spathodea campanulata)*

Be-still-tree *(Thevetia peruviana)*

Oleander Hedges Decorate the Islands

Oleander, *Nerium indicum,* are common landscape shrubs. The plants are known to be toxic, so be careful not to use the branches for any purpose.   Also called Yellow Oleander, the Be-Still-Tree gets its name from the constant movement of the leaves in the gentlest of breezes.

Hong Kong "Orchid," *Bauhinia blakeana,* resembles an orchid flower, but is in the legume family of plants. Orchid flower petals come in whorls of three; the Hong Kong "Orchid" has five petals, commonly found in legume flowers.

Island of Kauai

Being the oldest of the four most traveled islands, it is also the wettest spot on earth with a 460 inch average rainfall per year on the top of Mount Waialeale. (Wai in Hawaiian means "water.") Streams and waterfalls bring an abundant resource of water to the valley.

Kauai is a major producer of Taro (*Colocasia esculenta)*, a native food crop used to make poi. (We like poi! It is really good eaten with lomi lomi salmon, poke, and kalua pork at the luaus.)

Used as agricultural land for generations, the fields are beginning to produce crops such as coffee and macadamia nuts. Pineapple and sugar cane harvests are in decline on this island.

Known as the Garden Island, Kauai is recovering beautifully from the damage of Hurricane Iniki.

Currently, the entire Hanalei Valley is under the jurisdiction of the United States Department of the Interior as a Wildlife Refuge to preserve endangered birds and is not open to the public. Privately owned, this taro farm is within the Wildlife refuge. The taro fields have proven to be a suitable habitat for these birds.

Observed from the Princeville lookout on the road to Hanalei, this valley imparts a timelessness as mountain and man and woman work together.

Hanalei Valley and Mount Waialeale

National Tropical Botanical Garden (NTBG)

Known as the **Pacific Tropical Botanical Garden** until 1993, the name was changed to accommodate the addition of **Kampong Garden** in Coconut Grove, Florida. Four gardens are in Hawaii: **Allerton Garden** at the oceanfront in Lawai Valley on the south shore of Kauai; **Lawai Garden** adjacent to and mauka (towards the mountains) in Lawai Valley; **Limahuli Garden** in Limahuli Valley at Haena on the north shore of Kauai (page 69); and **Kahanu Garden** at Hana on the island of Maui (page 92).

Bamboo (Phyllostachys sp.)

Allerton Garden

A garden designed with elegant outdoor rooms, this place may feel like a grand, roof-less mansion that encompasses us with living walls. Water features are seen or heard throughout the grounds. Pause to look back, for the perspective of the garden changes.

On Highway 520, drive to Koloa. At Koloa take Poipu Road to the next fork, keeping right on Lawai Road following signs to Spouting Horn. NTBG's visitor center and parking lot is on the right. Reservations are required for all tours. Call ahead (808-742-2623) to arrange a time. Admission fee is charged.

Tillandsia tricolor

Hillsides and Steep Banks Covered with Bougainvillea

Song of India, *Dracaena (= Pleomele) reflexa*, in bloom

Queen Emma Spider Lily *(Crinum augustum)*

Lawai Stream, Where Ocean and Fresh Water Mix

Cigar Flower, *Calathea lutea*, is sometimes mistaken for a ginger because of the large spindle or "cigar-shaped" inflorescense.

Snake Plant, *Sansevieria trifasciata*, as ground cover

Climbing Pandanus, *Freycinetia multiflora*, with its peach colored flowers, doesn't seem to resemble the Pandanus family until the tiny "pineapple" shaped fruits are examined. The vining habit of the plant makes it an attractive cover for stone fences and other rough landscape areas.

Koki'o *(Hibiscus kokio)*

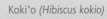

Dogbone Tree, *Polyscias nodosa*; see page 11

Staghorn Fern *(Platycerium bifurcatum)*

Moreton Bay Fig Trees (Ficus macrophylla) in Allerton Garden

Lawai Valley, Home of NTBG's Erythrina Collection

Lawai Stream

Lawai Garden

Headquarters for the NTBG are located here. This valley garden promotes ongoing scientific study and research while preserving its native beauty.

Hidden Lily *(Curcuma sp.)*

66

Waihulili Waterfall & Pool on Lawai Stream, Lawai Garden

Waihulili waterfall and pond are at the upper end of the valley. Lawai Stream flows down into the ocean. This natural wedge of land is called ahupua'a in Hawaiian, indicating a specific territory which supported an extended family.

Wild Impatiens mask an old stone wall just a few yards away from the pond.

Vahana Palm *(Pelagodoxa henryana)*

'Ilima *(Sida fallax)*

Epiphytic Cactus *(Schlumbergera sp.)*

Ko'oloa 'ula *(Abutilon menziesii)*

Otaheite or Gooseberry Tree *(Phyllanthus acidus)*
The fruits can be used for preserves.

Limahuli Garden

Ancient terraces in this part of the valley command our attention. Some are planted with Taro, *Colocasia sp.*, and show how the fields are irrigated. These restorations permit Limahuli Garden to pride itself as being "A Window To Ancient Hawaii" as stated on their trail guide booklet.

Here at Limahuli, conservation of native species is a priority while introduced plant material is being removed. Introduced plants including Guava, Mango, Octopus Tree, and the Autograph Tree all crowd out and overcome native plants.

Mass and might of the mountains become part of the Limahuli experience. Standing at the visitor center, we can see sheer cliffs rise up to the sky, as towering fortress walls. A hike onto the mountainside above the garden reveals spectacular views.

Ancient Taro Terraces

View from the Mountainside

Koki'o 'ula or St. John's Hibiscus, *Hibiscus kokio* ssp. *St. Johnianus,* is native to Kauai.

Alula, *Brighamia insignis,* sometimes described as looking like a cabbage on a baseball bat, is native to Kauai and Niihau.   Another native plant, the Ae'ae, serves as a ground cover surrounding the Alula.

Makana Mountain

Limahuli Stream

Mountain Apple, *Syzygium malaccense*, produces a sweet, edible pear-shaped fruit. Spent blossoms fall to the ground to produce a fine red carpet.

Commercial production of Guava fruits *(Psidium guajava)*

Guava Kai
Plantation

Drive northward on Highway 56, coming to Kilauea. Watch for Kuawa Road. Guava Kai Plantation signs and Guava orchards alongside the road direct us to the visitor center. Guava jams, jellies, syrups and sauces as well as other tropical fruit products are available here. Take a walk through their interesting collection of plants.

Cotton-Rose *(Hibiscus mutabilis)*

Dwarf Poinciana *(Caesalpinia pulcherrima)*

Surinam Cherry *(Eugenia uniflora)*

Olu Pua Gardens
and Plantation

Ice Blue Calathea (Calathea burle-marxii)

Hoop Pine, *Araucaria cunninghamii,* (center);
Norfolk Island Pine, *A. heterophylla* (right).

The Norfolk Island Pines on the right show damage after Hurricane Iniki stripped the branches, leaving the trees looking like utility poles. They are regaining growth, but some still look spindly.

Red Powderpuff (Calliandra haematyocephala)

Olu Pua is a Hawaiian phrase meaning "at peace among the flowers" or "floral serenity." Founders of Kauai's largest pineapple plantation established a twelve-acre garden estate. Today this historic plantation residence and magnificent garden are included in a one hour guided tour.

Admission fee is charged. Drive one mile west of Kalaheo on Highway 50. Plan to visit Olu Pua on the way to Waimea Canyon.

Elephant Apple *(Dillenia indica)*

Dwarf Malay Coconut Palm *(Cocos nucifera)*

Lollipop Plant *(Pachystachs lutea)*

Cuban Pink Trumpet *(Tabebuia pallida)*

Nun's Orchid *(Phaius tankervilliae)*

White Bauhinia *(Bauhinia variegata 'Candida')*

Poipu area was "target center" when Hurricane Iniki came ashore in 1992. As recovery from this devastation has moved along, Kauai's lovely gardens have also been restored.

Perhaps no other garden has been more carefully restored: Moir Garden at Kiahuna Plantation Resort is a "cactus" collection! Please don't let a few thorns thwart a delightful, if not cautious, "spine-tingling" adventure.

In addition to true cactus, there is more to meet the eye. Other succulents are found here: Euphorbias, Aloes, and Crassulas.

Lovely water lily ponds make a stark contrast to the arid garden.

Tropical Water Lilies (Nymphaea spp.)

Moir Garden

Candelabra Tree, Euphorbia ingens, at Kiahuna Plantation Resort

Red Aloe *(Aloe vera)*

Ghost Plant *(Graptopetalum paraguayense)*

Even though it may have a palm-like appearance, the Pony Tail Palm, *Beaucarnea recurvata*, is in the Agave family.

Felt-bush *(Kalanchoe beharensis)*

Opuntia cactus *(Opuntia sp.)*

Kalalau Lookout, more than 4,000 feet above Sea Level!

Waimea Canyon Drive and Kokee State Park

Nene Goose, Hawaii State Bird

'A'ali'i *(Dodonaea eriocarpa)*

White Rose *(Rosa sp.)*

Koai'a *(Acacia koaia)*

Although there are no formal garden settings along Highway 550 (Kokee Road) or Waimea Canyon Road, these routes will show plants growing in natural habitat. Some, like Grevillea (which is in the Protea family), are unique when in full bloom.

Driving to the end of the road is a must, even though there is a possibility that the Kalalau Lookout may be clouded over. We have it on good authority that the best time to be at the lookout is early in the morning before the cloud banks close in.

Watch for Nene Geese there.

The Kokee Museum shows videos of Hurricane Iniki and provides information about the Islands. It houses interesting artifacts of ancient culture and "hands on" exhibits. Gifts, books, ukuleles, and souvenirs are offered in this quaint museum's shop.

Silk Oak Tree *(Grevillea robusta)*

Onamental Banana *(Musa velutina)*

Red Iresine, *Iresine herbstii*, mimics a lava flow in this Coconut Plantation shopping center complex.

Garden Patches on Kauai

Autograph Tree, *Clusea rosea*, with Rose-like flowers

Gold Tree, *Tabebuia donnell-smithii*, (above) and Scrambled Eggs, *Cassia surattensis*, (below) near Spouting Horn.

The hard leaves are traditionally used to leave one's mark behind! Good used as landscape specimens, in the wild they become weeds, literally choking out native species as does the strangler fig.

Nene Geese, Hawaii's State Bird,
on Haleakala Mountain
with Maui Valley Below

Island of Maui

Hana Road is a Drive Through a Natural Garden

Sunrise Protea Farm

At the uplands of East Maui, soil conditions and climate are nearly ideal for growing Proteas ("PRO-tee-us"). On the cool slopes of Haleakala Mountain, these plants thrive at elevations of two- to four-thousand feet.

Native Protea habitat is found mainly in Australia and South Africa. They are not tropical plants but certainly fit into the realm of exotic flowers.

Pincushion Protea, Leucospermum cordifolium, Come in Many Colors

The Crowning Glory of a King Protea

A Grand Opening of King Protea
(Protea cynaroides) Inflorescence

The Crowning Glory of a King Protea

Scarlet Bottlebrush *(Mimetes cucullatus)*

Silver Tree *(Leucadendron argenteum)*

Golden Banksia *(Banksia prionotes)*

Pink Mink Protea *(Protea neriifolia 'May Day')*

Tree Banksia *(Banksia sp.)*

Kula Botanical Garden

Color Accents Highlight the Garden Walk

Agave, *Agave attenuata*, flowering

Cup of Gold Flower, sometimes called Goblet Flower, has a unique fragrance that reminds some people of coconut and others of vanilla ice cream. Give it your own sniff test.

Cup of Gold *(Solandra maxima)*

Rosemary Grevillea *(Grevillea rosemarinifolia)*

South African Heather *(Erica persoluta)*

Brazilian Snapdragon *(Ottocanthus sp.)*

Black-eyed Susan Vine *(Thunbergia alata)*

Kula Botanical Garden is compact and delightful! Get close to a Kangaroo Paw (page 90), meet the Koi, and check out a wall ablaze with Black-eyed Susan Vine. Being on a cool slope of Haleakala, this garden snuggles into the rocky mountainside.

Ice Plant *(Lampranthus sp.)*

Enchanting Floral Garden of Kula, Maui

Ice Plant *(Lampranthus multiradiatus)*

Upcountry Kula is cooler, being at a higher elevation. Breezes freshen, bringing winter-weary travelers an essence of springtime. This garden is a good place to take a break on the drive going to or coming from Haleakala Crater.

More than five hundred species of plants are growing here. As we follow along the pathways, our senses are treated to living canvasses of colors; fragrances and bird songs are carried on the gentle winds.

Flowering Maple *(Abutilon striatum 'Thompsonii')*

Trailing Abutilon *(Abutilon megapotamicum)*

Green Jade *(Strongylodon macrobotrys)*

White Powderpuff *(Calliandra portoricensis)*

Crown Flower *(Calotropis gigantea)*

Thunbergia *(Thunbergia mysorensis)*

Trailing African Daisy *(Osteospermum fruiticosum)*

Shower Orchid *(Congea tomentosa)*

Windmill Jasmine *(Jasminum nitidum)*

Scarlet Plume *(Euphorbia fulgens)*

Chinese Hat Plant *(Holmskioldia sanguinea)*

Baby Sun Rose *(Aptnia cordifolia)*

Air Plant *(Kalenchoe pinnata)*

Treasure Flower *(Gazania nevea)*

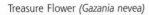

Rock Rose *(Cistus sp.)*

Bower Plant *(Pandorea jasminoides)*

Lavender Star Bush *(Grewia caffra)*

Hybrid Mandevilla *(Mandevilla x amabilis 'Alice')*

Pomegranate
(Punica granatum)

Pedilanthus *(Pedilanthus bracteatus)*

Red Kangaroo Paw *(Anigozanthos flavida)*

Bougainvillea Hedge, Enchanting Floral Garden of Kula, Maui

On the road to Hana, give your driver a break. Park at the gate and enjoy a walk back into this ravine arboretum. Groves of yellow and green bamboo fill the stream bank. Groupings of Torch Ginger polka-dot the view with their vivid color. Look for a stand of Mindanao Gum Trees with their exotic multicolored trunks.

Further back, ancient taro patches are being maintained. Now and again, these patches are raided by wild pigs feasting on the taro.

Kukui Nut Trees, various Palms such as the Coconut, Bottle, and Fishtail Palm, Mountain Apple Trees and Banana Plants grow here.

Mountain Apple *(Syzygium malaccense)*

Ti *(Cordyline terminalis)*, flowering

Keanae Arboretum

Hillside of Impatiens (Impatiens sp.) Growing Wild

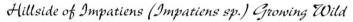

91

Kahanu Garden

At the edge of the land, where the ocean scrubs the shore with sand, this garden begins. Important for its agricultural capability, this site also preserves the Pi'ilanihale Heiau and the graves of the Hawaiian families who owned this property.

One of the National Tropical Botanical Gardens, Kahanu is known for its extensive collection of Breadfruit Trees, currently under scientific study.

This ahupua'a is a major historic, cultural, and research site. Please call (808-248-8912) to schedule a visit.

Breadfruit *(Artocarpus altilis)*

Restored Wall at Kahanu Garden Archeological Site

'Awa, *Piper methysticum;* made into a drink that has sedative properties, it helps induce sleep. It is a member of the black pepper family of plants.

Pandanus, *Pandanus sp.,* (below) is unusual among gardens in the Islands. Gray-green leaves and round tightly clustered fruits are a contrast to the common *Pandanus tectorius* silhouetted in the ocean view (opposite).

Beach Morning Glory *(Ipomoea pes-caprae)*

Ocean View from Kahanu Garden

Beach Heliotrope
(Messerschmidtia argentea)

Windmill and Landscape at Maui Tropical Plantation

Bat Flower *(Tacca chantrieri)*

Zig-Zag Plant *(Pedilanthus tithymaloides)*

Crown Of Thorns *(Euphorbia milii)*

Allamanda *(Allamanda cathartica 'Hendersonii')*

Quesnelia Bromeliad *(Quesnelia arvensis)*

Maui
Tropical
Plantation

At the pond: Angels Trumpet *(Brugmansia x candida)*, Bougainvillea *(Bougainvillea sp.)*, Canna *(Canna x generalis)*, Kukui NutTree *(Aleurites moluccana)*, Red Ti *(Cordyline terminalis)* paint the landscape.

Orange-red Popcorn Vine, *Norantea guianensis,* and blue spikes of Pickerel Weed, *Pontederia lanceolata,* accent this pond with their showy blossoms.

96

Large clusters of bracts weight down the tall stems of the Red Ginger, *Alpinia purpurata*. The single cluster of the cultivar 'Jungle King' (right) shows tiny white flowers protruding between the red bracts. These flowers do not produce seeds that drop from the plant, but new plant "babies" develop in place producing the heavy plume-like clusters seen above.

Red Ginger *(Alpinia purpurata 'Jungle King')*

Tropical Gardens of Maui

Located beyond Wailuku, the Ioa Valley Road is an adventure all its own. Sheer cliffs, like great fortress walls, confine us; mountains confront us; valley beauty surrounds us.

After a rainstorm, waterfalls shoot down the mountains like liquid arrows. The Iao Stream seems to boil in its courseway to the sea.

Located beyond Wailuku on Iao Valley Road, the gardens are established on both banks of the Iao Stream. Please go slowly through these gardens. The landscapes touch the soul as quickly as they please the eye.

Green Ice Calathea
(Calathea burle-marxii 'Green Ice')

King Ixora *(Ixora macrothyrsa)*

Aechmea Bromeliad *(Aechmea sp.)*

Tropical flower mix for floral arrangements

Song of India, *Dracaena (= Pleomele) reflexa* 'Variegata'

Parrot Beak Heliconia *(Heliconia psittacorum)*

Angel's Trumpet *(Brugmansia x candida)*

Koi Pond, Tropical Gardens of Maui

Garden Patches on Maui

Silversword, Argyroxiphium sandwicense ssp. macrocephalum, Endemic to Haleakala

Silversword is endemic to the Hawaiian Islands, emerging after centuries of enduring harsh conditions at high altitudes. Other species of *Argyroxiphium* have developed on Mauna Kea and Mauna Loa on the Island of Hawaii and also on the Island of Kauai.

Look for the Silversword garden at the Ranger station (7,000 feet elevation) on Haleakala.

Commercial Pineapple
(Ananas comosus)

Princes' Vine
(Ipomoea horsfalliae)

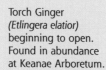

Torch Ginger
(Etlingera elatior)
beginning to open.
Found in abundance
at Keanae Arboretum.

Half-flower (Scaevola frutescens) at Oheo Gulch, Haleakala National Park

Half-fower *(Scaevola frutescens)*

Beach Morning Glory *(Ipomoea pes-caprae)*

Yellow Vein
(Pseuderanthemum reticulatum)

Stephanotis *(Marsdenia floribunda)*

Stephanotis fruit

Stephanotis Vine covers fences throughout the Islands, providing fragrance and privacy. It is in the milkweed *(Asclepediaceae)* family as evidenced by large follicles that break open when ripe to reveal equally large, tufted wind-catching seeds.

Wind-catching seeds blow away, landing in places where they can begin again. Garden journeys also bring us to places where we will, in time, begin again.

Stephanotis ripe fruit and seed

Ti Basket with Plumeria Blossoms

Aloha!

Bibliography

Baensch, U. and U. Baensch. 1994. Blooming Bromeliads. Tropical Beauty Publishers.

Bailey, L. H. Hortorium Staff. 1976. Hortus Third. Macmillan.

Bostwick, J., D. Peebles and A. Kepler. 1987. Pua Nani. Mutual Publishing.

Courtright, G. 1988. Tropicals. Timber Press.

Engebretson, G. 1993. Parker Ranch. Legacy Publishing.

Graf, A. B. 1978. Tropica. Roehrs.

Kepler, A. K. 1989. Exotic Tropicals of Hawaii. Mutual Publishing.

Kepler, A. K. 1995. Maui's Floral Splendor. Mutual Publishing.

Kepler, A. K. 1988. Proteas in Hawaii. Mutual Publishing.

Kepler, A. K. 1990. Trees of Hawaii. University of Hawaii Press.

Krauss, B. H. and T. Greig. 1993. Plants in Hawaiian Culture. University of Hawaii Press.

Lamoureux, C. 1976. Trailside Plants of Hawaii's National Parks. Hawaii Natural History Association.

Miyano, L. and D. Peebles. 1995. Hawai'i a Floral Paradise. Mutual Publishing.

Sohmer, S. H. and R. Gustafson. 1987. Plants and Flowers of Hawaii. University of Hawaii Press.

Wall, B. and C. Innes. 1994. Air Plants and other Bromeliads. Cassel Educational Limited for Royal Horticultural Society.

Warren, W. and L. I. Tettoni. 1997. The Tropical Garden. New edition. Thames and Hudson.

Whistler, A. 1980. Coastal Flowers of the Tropical Pacific. Oriental Publishing Co. for The National (Pacific) Tropical Botanical Garden.

Numerous brochures, pamphlets and trail guides from the botanical gardens are useful sources of information.

Family Fun!
Plant Families and Genera

(Common family names in parentheses)

* Botanical names are goverend by the International Code of Botanical Nomenclature. Family names are usually recognized by the suffix aceae attached to the stem of one plant genus within the family. There are eight families that were historically formed in another manner those names have been used so long that they are traditionally accepted. The five indicated with an (*) have both the name that follows the International Code rules and the traditional name listed.

*Kepaniwai Heritage Gardens in the Iao Valley at Wailuku,
Island of Maui, with gardens representing major ethnic
groups that have come to the Hawaiian Islands*